ISBN 978-0-364-37650-8
PIBN 11278719

1 MONTH OF
FREE
READING

at
www.ForgottenBooks.com

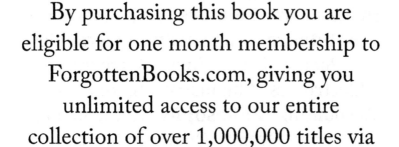

By purchasing this book you are eligible for one month membership to ForgottenBooks.com, giving you unlimited access to our entire collection of over 1,000,000 titles via our web site and mobile apps.

To claim your free month visit:

www.forgottenbooks.com/free1278719

English
Français
Deutsche
Italiano
Español
Português

www.forgottenbooks.com

Mythology Photography **Fiction**
Fishing Christianity **Art** Cooking
Essays Buddhism Freemasonry
Medicine **Biology** Music **Ancient
Egypt** Evolution Carpentry Physics
Dance Geology **Mathematics** Fitness
Shakespeare **Folklore** Yoga Marketing
Confidence Immortality Biographies
Poetry **Psychology** Witchcraft
Electronics Chemistry History **Law**
Accounting **Philosophy** Anthropology
Alchemy Drama Quantum Mechanics
Atheism Sexual Health **Ancient History**
Entrepreneurship Languages Sport
Paleontology Needlework Islam
Metaphysics Investment Archaeology
Parenting Statistics Criminology
Motivational

MERA PUBLICATIONS.

No. 1.

The Scientific and Economic Aspects of the Cornish Pilchard Fishery.

1.—The Food and Feeding Habits of the Pilchard in Coastal Waters.

By HAROLD SWITHINBANK,

Fellow of the Royal Society of Edinburgh,

and G. E. BULLEN.

ST. ALBANS PRESS,

1913.

FOREWORD.

In placing the following paper before the notice of those interested in Marine Biology the writers desire it to be understood that the present contribution forms the first of a series of pamphlets to appear at irregular intervals. The subjects to be dealt with relate largely to the scientific and economic aspect of certain fisheries, together with phases of Marine Bionomics having more or less connection therewith.

Applications for copies of the present paper may be addressed to G. E. BULLEN, The Hertfordshire Museum, St. Albans.

MERA PUBLICATIONS.

No. 1

The Scientific and Economic Aspects of the Cornish Pilchard Fishery.

1.—The Food and Feeding Habits of the Pilchard in Coastal Waters.

By HAROLD SWITHINBANK,

Fellow of the Royal Society of Edinburgh,

and G. E. BULLEN.

ST. ALBANS PRESS.

1913.

The Scientific and Economic Aspects of the Cornish Pilchard Fishery.

I.—The Food and Feeding Habits of the Pilchard in Coastal Waters.

By Harold Swithinbank, f.r.s.e., and G. E. Bullen.

The researches here described are founded upon two distinct series of observations.

The first were obtained at intermittent periods during the years 1905-07, and embraced enquiry into the food problem of herring, mackerel, and pilchard, frequenting the western part of the English Channel. To these are added the results obtained from a collection of stomach material made by fishermen in Mount's Bay in 1906, and a similar series from Mevagissey Bay supplied by Mr. Howard Dunn in 1907.

The second group of observations were obtained in 1913 during a planktonic and hydrographical survey of the principal stations affected by pilchard on the Cornish seaboard, performed from the s.s. Mera R.Y.S.

In view of the fact that no exact determination of the nature of the food of pilchards in English waters has ever been published, the following paper is put forward as a preliminary step toward a deeper consideration of the food supply in relation to migration.

The authors' thanks are due to the Marine Biological Association of the United Kingdom, under the auspices of which the first series of observations were obtained; also to Mr. Howard Dunn of Mevagissey and Mr. Matthias Dunn of Newlyn, for material and observations supplied.

On the French, Portuguese and Mediterranean seaboard, the food and habits of the sardine, or pilchard, in its early and

later stages of growth has received some considerable attention at the hands of such investigators as Marion,[1] Gourret,[2]
Pouchet and de Guerné,[3] Canu,[4] H. M. Smith,[5] and Casimir
Cépède.[6] The work of this last named author is particularly
precise, showing as it does that numerous zoo- and phytoplankton organisms are to be met with in stomach contents,
and that the species described are common to the plankton
of the area from whence the fish are derived. Unfortunately,
however, no particulars are given as to the sizes of the
individual fish examined. Nor is there any evidence put forward to show whether this investigator was enabled to work
out stomach material to the same detailed extent as a plankton
sample, a somewhat important point where the question of a
selective capability on the part of the fish itself has to be
considered.

The same author remarks upon the finding of the flagellate
Phæocystis poucheti (Har.) in large quantity in the stomach
contents of fish examined in April, 1906, a particular form of
food which may be referred to more conveniently further on.
Generally speaking, however, M. Casimir Cépède's papers
show conclusively that plankton formed the sole food of the
fish coming under his observation at all seasons.

The published information respecting the food of the pilchard in English waters is, as previously inferred, of too vague
a description to be of material use in the consideration of the
food problem from a wider aspect. Yarrell[7] states: "They

[1] Marion, A. F., "Remarques générales sur le régime de la faune
pélagique du golfe de Marseille particulièrement durant l'annee 1890" (Ann.
du Mus. de Marseille, p. 128, etc.)

[2] Gourret, Paul, "Examen de la pâture de quelque poissons comestibles
du golfe de Marseille" (*id.*, 1890, p. 30).

[3] Pouchet et de Guerné, "Sur la nourriture de la Sardine" (comptes
rendus, t. civ., 1887, p. 712-715).

[4] Canu, E., "Notes de biologie marine faunique ou èthologiques iv.
Diatomèes et Algues pèlagiques abondantes dans la Manche du N.E."
p. 113-116 (Ann. de la Stat. Aquic. de Boulogne sur Mer, T. 12th part.,
Juin, 1899).

[5] Smith, H. M., "The French Sardine Industry" (Bull, U. S., Fish.,
Com. xxi., 1901, p. 1).

[6] Cépède, M. Casimir, "Quelque remarques sur la nourriture de la Sardine," 1906, and "Contribution à l'ètude de la nourriture de la Sardine," 1907.

[7] Yarrell, Brit. Fishes, vol. ii., p. 98.

feed with voracity upon small crustaceous animals, and I have found their stomachs crammed, each with thousands of a minute species of shrimp not larger than a flea." Couch,[8] somewhat more precise, states that "at the end of July the food is supposed to be vegetable in character; the examination of stomachs revealed vegetable matter. In mid-summer crustaceans are largely taken at the surface." The same author goes on to say : "On one occasion stomachs of several were found to contain mackerel midge"; the roe of cod and ling is also given as an article of diet. Day[9] describes the food as being similar to that of herring, and probably, in common with Couch, on the strength of his knowledge of methods employed in the French sardine fishery (pêcherie de rouge), states that the pilchard voraciously devours "the roe of cod and ling, also bread crumbs." The late Mr. Matthias Dunn, whose observations upon the natural history and habits of the species in Cornish waters afford much valuable information, gives in one of his papers[10] a more exact account of the food derived from an examination of stomach material. In the case of fish *presumably* arriving from deep water the stomachs were found to contain large numbers of the Copepod *Anomalocera pattersoni* and *Thysanopoda* (*Nictyphanes couchii*). Copepods and Amphipods were stated to be the Atlantic food of the species, whilst zoeas and the "spores of the olive seaweed" were considered by this author to serve largely as the food in inshore waters. A particular instance moreover of a fish feeding upon prey of a large character is also given[11] in the same paper, the author stating that on Sept. 3rd, 1889, he examined an example which had in its stomach forty young fish of three distinct species. Easily distinguishable were the Crystal Gobies, plentiful off the Deadman, and young Launces; the third species seemed very much like young pilchards.

Some comment upon the above observations at this point

---

8 Couch, J., " Fish of British Isles," vol. iv., p. 83.

9 Day, " British Fishes," vol. ii., p. 225.

10 Dunn, M., " Migrations and Habits of the Pilchards," Lectures on Fishes, etc., County Fisheries Exhibition, Truro, 1893, p. 159, etc.

11 *id.*, p. 174.

may be desirable.   In the first place it may be remarked that the occurrence of such a Copepod as *Anomalocera pattersoni* R. Temp. and the Euphausid *Nictyphanes couchii* Bell in stomach contents cannot, in the absence of plankton observations taken in the area immediately on, or adjacent to the fishing grounds, be accepted as evidence that the fish had come in from deep water, since inshore visitations of such plankton species have been recorded.   From a careful consideration of Mr. Dunn's observations it is evident that considerable importance as an article of diet is attached to material described as "the spores of the olive seaweed."   The present authors have been to some considerable pains to determine the true nature of this organism, but without success.   From information derived from Mr. Howard Dunn, of Mevagissey, it would appear that in former years this appeared in large quantity in Mevagissey and St. Austell's Bays about the beginning of June, its paucity or abundance apparently affecting the extent of inshore migration of pilchard, and that the fish appeared to feed ravenously upon it until the supply was exhausted.   Of recent years this organism has not appeared, and in the opinion of Mr. Dunn and certain other observers interrogated upon the subject, the decline of the fishery may to a certain extent be attributed thereto.

The only suggestion which may be hazarded as to the true nature of this organism is that it is either the Flagellate *Phaeocystis globosa* Scherfel or the Chlorophycean *Halosphaera viridis* Schmitz—in greater probability the former.

Numerous observations of the more or less abundant occurrence of *Phaeocystis* at the International Cruise Stations situated in the Great West Bay, at Plymouth, etc., have been recorded,[12] but unfortunately none of these positions lie within the usual pilchard fishery area.   During the early part of June, 1913, the present authors, whilst tow-netting in St. Ives Bay, met with a profusion of *Phaeocystis*, the same organism appearing to be absent from other stations investigated on the South Coast at the time.   From information derived from an

---

[12] *Vide,* Plankton tables Inter. Investigations, "Bulletins Conseil Inter. pour l'exploration de la Mer."

observer situated at Milford in August of the same year, *Phaeocystis* occurred in abundance along the coastal waters of South Wales and the adjacent area, whereas at this time from our personal observation it had entirely disappeared from St. Ives Bay. The periodic occurrence of this species has already been described by Dr. L. H. Gough,[13] but it may here be remarked that the estimation of its frequency, and in fact its correct determination when occurring in stomach material and preserved plankton samples, is at all times difficult.

M. Casimir Cépède's finding of *Phaeocystis poucheti* (Har.)[14] in the stomach contents of fish examined in the spring of 1906, agrees with our own observations made upon mackerel and pilchard in April, 1906, and March and April, 1907.

The available observations lending support to the theory that this particular type of food on the other hand may be *Halosphaera viridis* are fewer in number, and none of the Continental authors hitherto cited make mention of the occurrence of the species in material examined. In April, 1906, it was observed in more or less regular abundance in plankton collected on the Western mackerel grounds,[15] and was occasionally observed in the stomach contents of a few pilchards taken in the mackerel nets.

An examination of the plankton tables for those International Cruise Stations situated in the near vicinity to the pilchard fishery area shows that the species is of far less frequent occurrence than *Phaeocystis* in the first half of the year, and that it does not belong to the group of organisms known as " periodic plankton."

Passing now to a general consideration of the food of the pilchard, as determined from the observations obtained during the present enquiry, it may first be stated that the fish from

---

[13] L. H. Gough, "Rep. on the Plankton of the Eng. Channel," Inter. Inv. Mar. Biol. Ass. Rep. 1, p. 339.

[14] Note.—The determination of the species, occurring at the Inter. Cruise Stations was for some time incorrect (*vide* Gough, op. cit. p. 339), and it is possible that a similar error may here exist, since the identificaton was performed on stomach material.

[15] *Vide* Bullen, G. E., " Plankton Studies in relation to the Western Mackerel Fishery," Table No. 1.

which material was derived were the product of the ordinary drift fishery. The comparative irregularity of seining, etc., rendered the process of obtaining fish from other sources an uncertain one, and as the taking of small examples is only occasionally effected by indirect means, the food conditions to be described relate solely to the sexually unripe fish which form the subject of the Cornish fishery. The ordinary average length attained by such individuals does not appear to vary to any considerable extent, probably due to the fact that a moderately constant diameter of mesh is employed in their capture.

The following table compiled from a series of fish measured in August, 1913, shows the extent of length variation occurring in an ordinary catch. But it may here be

TABLE NO. I.

MOUNTS BAY, AUGUST 18TH—21ST, 1913.

TABLE SHOWING PERCENTAGES OF SIZES IN SUMMER FISH.

Length* in m.m.	Number of Fish.
208	1
209	2
211	1
215--220	5
221—225	7
226—230	11
231—235	14
236—240	17
241—245	13
246—250	17
251— 255	9
257	2
258	1
	Total 100 fish.

* Note.—Measurements in every case were taken from the tip of the snout to the extremity of the upper lobe of the tail.

remarked that fishermen and fish packers and curers, interrogated on the subject, agree in stating that the size of the summer pilchard has materially increased within the last decade, and that this condition is contemporaneous with the present decline in the fishery. One curer of forty years' experience seemed to sum up the general opinion in stating that twenty years previously fish were altogether more numerous, but inferior in size and quality to those of the present day.

An examination of the "Sea Fisheries Statistical Tables" serves to demonstrate the fact that although pilchards are marketed in appreciable quantity from July to January, the heavy landings take place from August to November. From a consideration of these facts it may be assumed that during these months inshore migration reaches a maximum. From the writers' personal enquiry amongst the fishing community at Plymouth, Mevagissey, Newlyn, and St. Ives, it would appear that in former years two distinct migrations took place during the year. The more important still forms the object of the summer fishery which is carried on throughout the inshore waters from Plymouth to St. Ives; the second, occurring in winter, gave rise to a fishery within a few miles radius of the Eddystone.

The figures previously referred to do not show total landings above 1,000 cwt. for December since the year 1902, a fact which serves to demonstrate that the winter fishery has been steadily on the decline during the past decade. At the present time it may be considered to be practically non-existent.

It is generally stated by fishermen that these winter pilchards were of considerably larger size and better quality than the summer fish. In the opinion of Mr. Howard Dunn of Mevagissey, Mr. Matthias Dunn of Newlyn, and other observers of extensive experience, they constituted either a separate race or were late spawners making an inshore migration after the advent of the sexually immature fish. A discussion upon this point is beyond the province of the present paper, but the above facts have been stated with a view to showing that in any consideration of the question of food we have two distinct types of fish to deal with.

In view of the paucity of our examinations made upon these winter fish, it will be convenient at this point to deal with certain observations taken on October 11th, 1905, off Portwrinkle on a small shoal which was known to be moving eastward toward the Eddystone grounds, in which area fishing was continued for some days until the disappearance of the fish. The men in charge of the boat from which the observations were made, agreed in stating that the fish presented the appearance of a typical winter shoal, and negatived the suggestion that any of the individuals composing the catch of 1,500 were late summer fish. The appearance of a large number of seabirds (gannets, gulls, etc.) gave an impression that the fish were more densely shoaling than they ultimately transpired to be, but catches of 5,000 down distributed amongst nearly forty vessels fishing over an area extending from Rame Head to Looe, enclosing practically the whole of Whitesand Bay, showed that the fish were swimming in that particular type of open order which, so it is generally stated by fishermen, is characteristic of a winter shoal. Measurements of 100 fish from the tip of the snout to the curve of the tail showed a length variation from 18 to 24 centimetres.

Plankton samples taken on the position where the nets were shot showed a large variety of species, the amount of phytoplankton being in preponderance over zooplankton. The former was chiefly composed of diatoms, *Rhizosolenia stoltefothii* being very plentiful, whilst the animal matter was represented largely by the Copepods *Pseudocalanus elongatus* and *Centropages typicus*, together with *Obelia nigra*, *Sagitta bipunctata*, *Oikopleura dioica*, etc.

Examination of a series of stomachs showed a food condition very similar to that to be observed at times in early spring mackerel from the Western area, viz., a coherent mass of material of a greenish colour, which upon critical examination proved to consist of large numbers of *Oikopleura dioica*, *Sagitta bipunctata*, and occasional Copepods—*Pseudocalanus elongatus*, *Corycaeus anglicus*, *Centropages typicus*, etc.— together with a large quantity of diatomaceous matter, and the " house " of *Oikopleura*.

It has been mentioned that the Medusa *Obelia nigra* occurred in the plankton samples, its abundance causing considerable " briming " in the water along the fleet of nets, a fact which caused the fishermen to consider a large catch improbable. It is an important consideration therefore to note that although particular search was made for the manubrium of the species in the stomach material, no evidence whatever of the existence of a single example was forthcoming. Another matter of interest lay in the very high proportion of *Oikopleura* in the stomach samples in comparison with that observed in the plankton. Generally speaking, however, the species identified in the stomachs were common to those of the tow-nettings, but the variety occurring in the former was by no means so great as that of the latter.

It is interesting to compare the food conditions of the above with those observed in two series of stomachs from fish obtained in Mevagissey Bay on the 20th of October, 1905, and Mounts Bay on the 23rd of the same month. The fishery at both stations represented, at the time, the final stages of the summer season. The few vessels still engaged were shooting comparatively close inshore and obtaining small catches of 500 down, the fish being undoubtedly the remnants of the summer shoals together with a fair sprinkling of "shirmers," *i.e.*, spent fish of a thin cadaverous appearance, generally above the normal size.

The food observed in the first series (Mevagissey) was of a character very similar to that already described in the case of fish from Portwrinkle, but it was possible to detect a higher proportion of Copepod remains attributable to the following species :—*Acartia clausi, Calanus finmarchicus, Centropages typicus, Pseudocalanus elongatus, Temora longicornis*, etc. *Oikopleura dioica*, however, again preponderated, and by reason of the presence of a large quantity of phytoplankton the stomach contents generally were of a greenish colour.

In the Mounts Bay series, however, the conditions were entirely different. In the six stomachs examined there was no indication of any phytoplankton. *Oikopleura* was not observed, and the whole of the stomach contents presented a reddish brown appearance, interrupted here and there by layers of a darker colour. The former material was found to consist almost entirely of Copepods, amongst which *Calanus finmarchicus* preponderated; the streaks of a darker tint were due to an almost pure sample of the Pteropod Mollusc *Limacina retroversa* Flem., shoals of which had been frequently met with during the summer of the same year.[16]

It may be remarked at this point that according to the late Mr. Matthias Dunn[17] "shirmers" appear all along the coast about a month or six weeks after the summer fishing has

---

[16] *Vide* " Plankton Tables," Inter. Fishery Bulletins.
[17] Dunn, M., " The Migrations and Habits of the Pilchards," Lectures on Fishes, etc., County Fisheries Exhibition, Truro, 1893, p. 171.

commenced, and that these fish do not migrate into deeper water in the late autumn but remain inshore until they die. According to the statement of many fishermen, individuals of this type sometimes compose 25 per cent. of a catch made in the late autumn, and are invariably to be met with in a true winter shoal. The present authors have been informed by several fish curers and others at Newlyn and Mevagissey that food is very seldom to be met with in the stomach of a " shirmer," and there seems to be a fairly general opinion that death is hastened by the fish's voluntary abstinence from nutrition. It is important therefore to state that during the present investigation the stomachs of numerous " shirmers " have been examined, and that there appears to be no foundation for this belief, a normal quantity of food having been observed in the majority. In the series already mentioned two of the six samples from Mevagissey Bay and three from those taken in Mounts Bay were derived from " shirmers."

A further attempt at observation on these winter shoals, on this occasion directly on the Eddystone ground, was made from a fishing vessel on the 31st of January, 1906, but proved unsuccessful, only six pilchards being taken in a small mixed catch, viz., three " shirmers," and three males with generative organs fairly well developed. Examination of the stomach contents showed food of a character entirely distinct from that of the Portwrinkle fish referred to above : there was no evidence whatever of any phytoplankton, the bulk of the material being composed of fish larvae and ova, probably those of herring, together with a slight addition of three species of Copepods, *Caradid larvae, Oikopleura dioica* and *Sagitta bipunctata.* The fish larvae and ova formed by far the greater proportion of the sample in each instance, the stomach being distended to some considerable extent by the amount of materials contained therein.

The comparison of these results with those obtained from a series of plankton tow-nettings show certain remarkable features. Fish ova were recorded as rare, only occurring in the oblique haul, and larvae were not observed in this netting or at the surface. Phytoplankton, consisting largely of two diatoms, *Lauderia borealis* and *Biddulphia mobiliensis,* preponderated over zooplankton, this latter being composed of certain copepods and other metazoa, certain only of which were observed in the stomach contents.

We thus have an instance of food observed uniformly throughout six stomach samples of a type widely dissimilar from that of the plankton occurring on the ground from whence the fish were taken.

Passing now to a consideration of the food of the summer pilchard, the reader's attention is directed to the tabulated results obtained from a group of observations made in Mevagissey Bay and Mounts Bay in July and August, 1906.

Here we have the results obtained from four series of stomach material, together with that of a plankton sample, taken in each case on the position from which the fish were derived. A fairly regular correlation is shown between the stomach material and plankton throughout the first three series, but in the fourth we see three entirely aberrant examples, in which food of the type described for certain of the winter fish again makes its appearance.

It should be remarked that for the purposes of the present table only those species forming the bulk of the stomach contents have been noted in the case of the plankton samples, but in the working *in extenso* of the latter in the fourth series no correlative feature was found to exist to account for the food condition as indicated. We thus have another instance of material occurring in stomach contents not common to the plankton environment.

The next table (Table No. 3) shows the proportions of plankton species observed in a series of thirty-six stomachs taken from fish caught throughout the height of the summer fishery in Mevagissey Bay and the adjacent waters in 1907. Each of these samples were worked *in extenso*, practically the whole of the stomach contents being submitted to a critical examination. Unfortunately no plankton samples were taken at the same time as the stomach series, but a general review of the table affords some interesting comparisons.

In the first place it will be seen that certain species well represented in one sample will be entirely lacking or but sparsely shown in others of the same series, *e.g., Centropages*

TABLE No. II.—TABLE SHOWING RELATION OF FOOD TO PLANKTON. SUMMER FISHERY, 1906.

	MEVAGISSEY BAY.												MOUNTS BAY.											
Dates.	JULY 26TH.					AUGUST 15TH.							JULY 24TH.						AUGUST 10TH.					
Positions.	S.W.×W. of Chapel Point 3 m.					N. N E of Chapel Point 3 m.							E. S E. of Wolf 10 miles.						S.E. of Mousehole 3 miles.					
S = Stomach sample. P = Plankton.	S	S	S	S	P	S	S	S	S	S	S	P	S	S	S	S	S	P	S	S	S	S	S	P
Sample Nos.	1	2	3	4	5	6	7	8	9	10	11	12	13	14	15	16	17	18	19	20	21	22	23	24
Calanus finmarchicus Gunn	—	+	r	+	+	+	r	r	r	r	+	+	c	c	+	+	+	c	+	+	rr	rr	—	+
Centropages typicus Kröyer	c	c	+	c	+	c	+	+	+	+	+	c	c	+	+	+	+	r	r	+	—	—	—	+
Pseudocalanus elongatus Boeck	+	+	+	+	+	+	+	+	+	+	r	+	r	+	r	+	+	+	+	+	rr	—	—	+
Temora longicornis O. F. Mull	c	c	c	—	c	r	—	+	+	r	r	+	—	+	r	r	—	+	+	+	+	—	—	+
Caradid larvae and Zoeas	c	c	+	r	+	—	—	—	r	r	—	+	c	c	+	c	c	c	+	r	+	—	—	+
Limacina retroversa. Flem	cc	r	r	r	rr	—	—	—	r	r	r	rr	r	r	r	rr	c	c	r	r	c	rf	rf	r
Sagitta bipunctata Quoy and Gaim	+	r	r	—	rr	—	—	—	r	r	r	+	+	r	r	r	r	r	r	r	—	—	—	r
Oikopleura dioica Fol	r	r	rr	r	rr	—	—	—	r	r	rr	rr	—	—	—	—	r	r	—	—	c	+	+	r
Phytoplankton as described	—	—	—	—	—	—	—	—	—	—	—	rr	—	—	—	—	—	rr	—	—	cc	cc	cc	rr

NOTE.—The Symbols of frequency adopted in the above and other tables are similar in value and use to those employed in the Reports of the Inter. Investigations, published by the Marine Biological Association.

TABLE No. III.—TABLE SHOWING THE OCCURRENCE OF CERTAIN PLANKTON ORGANISMS IN SERIES· OF STOMACHS TAKEN FROM FISH CAUGHT IN MEVAGISSEY BAY AND ADJACENT AREA THROUGHOUT THE HEIGHT OF THE SUMMER FISHERY 1907.

Positions.	N.E. × E. of Dodman 4 miles.				S.W. of Dodman 8 miles.				Between Dodman and Gribben.				Hemmick Bay.				Between the Dodman and Falmouth.				Mevagissey Bay.				Mevagissey Bay.				Mevagissey Bay.				Mevagissey Bay.			
Dates.	July 8th-14th				July 15th-20th				July 24th-28th				Aug. 3rd-7th				Aug. 7th-12th				Aug. 13th-18th				Aug. 18th-22nd				Aug. 23rd-27th				Aug 28th-Sept. 3rd			
	A				B				C				D				E				F				G				H				I			
Sample Numbers	1	2	3	4	5	6	7	8	9	10	11	12	13	14	15	16	17	18	19	20	21	22	23	24	25	26	27	28	29	30	31	32	33	34	35	36
Acartia clausi Giesbr.	rr	r	rr	r	r	r	rr	–	–	r	rr	rr	+	r	r	r	–	–	rr	rr	–	r	–	–	rr	r	r	rr	rr	–	–	r	rr	r	–	rr
Calanus finmarchicus Gunn.	+	+	c	c	r	r	r	rr	r	r	r	r	r	r	r	r	r	r	rr	rr	r	–	rr	rr	rr	rr	r	rr	rr	r	rr	r	rr	rr	rr	–
Centropages typicus Kröyer	r	c	c	+	+	r	r	rr	r	r	r	r	–	+	r	r	+	+	rr	rr	rr	–	rr	rr	rr	r	r	r	rr	r	rr	rr	rr	rr	rr	rr
Corycaeus anglicus Lubb.	–	rr	–	rr	rr	r	r	rr	–	rr	–	r	–	r	r	r	r	+	rr	–	r	rr	rr	–	–	r	–	–	–	r	rr	r	rr	rr	–	rr
Paracalanus parvus Claus	+	+	+	+	+	+	+	+	+	+	+	+	+	+	+	+	+	+	+	+	+	+	+	+	+	+	+	+	+	+	r	r	r	+	+	r
Pseudocalanus elongatus Boeck	+	+	c	c	+	+	+	+	r	r	c	+	c	c	c	c	r	r	r	r	c	c	c	r	c	c	c	c	c	r	cc	r	c	c	cc	+
Temora longicornis O.F Mull.	+	+	c	+	+	r	r	r	r	–	+	r	+	+	c	+	r	r	r	r	+	+	–	r	r	r	r	r	r	r	+	r	c	c	c	+
Caradid larvae	r	–	rr	–	c	–	rr	r	–	–	–	–	r	+	–	–	rr	rr	r	r	cc	c	–	–	r	r	rr	c	rr	r	r	r	rr	r	r	–
Evadne Nordmanni Loven	r	–	–	–	r	–	–	–	–	r	–	r	r	–	rr	r	r	rr	r	r	r	–	r	rr	rr	–	rr	r	rr	rr	rr	r	–	–	r	rr
Podon intermedius Lilljeb	rr	–	–	rr	+	rr	–	rr	–	rr	rr	–	–	+	rr	r	r	r	rr	rr	+	r	r	r	rr	r	r	r	rr	r	r	r	rr	rr	r	r
Zoeas	rr	r	–	–	r	r	+	r	–	–	rr	–	rr	r	r	r	–	r	r	–	–	r	rr	+	–	r	r	r	rr	r	r	r	r	rr	r	r
Sagitta bipunctata Quoy & Gaim	r	–	rr	r	c	r	+	rr	–	–	r	rr	–	+	r	r	r	rr	rr	rr	+	+	r	–	rr	–	rr	rr	rr	rr	rr	–	+	–	r	–
Oikopleura dioica Fol.	–	rr	r	–	–	–	rr	r	r	–	–	–	rr	rr	r	r	–	–	rr	r	–	+	r	–	–	rr	r	r	rr	–	–	rr	–	–	–	–

Also observed all rr. or r.

Clytemnestra rostrata S. 2, 3,
Euterpe acutifrons, S. 2, 8,
Oithona similis S. 3, 7, 8, 9.
 [26, 28, 29, 30,
Cyphonautes, 1, 2, 4, 18, 20.
Echinopluteus, 7, 30, 32.
Cirripede larvae, 4, 9, 17.

typicus in A. D. E., etc.; *Caradid larvae* in B. F. G., etc. Again, whilst certain species, notably *Pseudocalanus elongatus* and *Temora longicornis*, sustain a more or less regular occurrence throughout the whole range of series, others are of more or less fitful or uncertain appearance. The most important point, however, appears to lie in the fact that the bulk of the food in every instance appears to have been composed of these two species of Copepods, which are undoubtedly rich in oil, and probably of high nutrient value.

It is important to note that, according to the " Fish Trades Gazette," the fishery in Mevagissey Bay throughout the period over which these samples were taken was of a regular and productive character; the official returns show the following total landings for the West Country.

1907	July	Aug.	Sept.	Oct.
South Coast .	2,285	29,639	40,464	5,140 cwts.
North Coast .	31	7,840	3,225	186 cwts

These figures do not represent a phenomenally good season, but demonstrate the fact that the fishery gradually improved from its commencement until it reached its height, and that the inshore migration was in accordance with what we may term normal[18] conditions.

In striking contrast to the foregoing series of observations are those shown in the next table (Table IV.), which represent the character of the food observed in a more extensive collection of stomachs obtained from the Mounts Bay and Mevagissey Bay fisheries in August of the present year (1913).

The nature of the plankton environment occurring at these two stations will be dealt with in detail in a subsequent paper, but it is desirable at this point to state that throughout the whole of Mounts Bay phytoplankton occurred in considerable preponderance over zooplankton. At certain positions lying off the Lizard the tow-nets, after a short immersion in the water, became choked with diatoms and gave a slimy impression to

---

[18] Note.—In view of the fact that the official returns show occasional difference in those months in which the heaviest landings have taken place throughout a number of years, the term "normal" is used to designate the condition exhibited by the majority.

TABLE No. IV.—TABLE SHOWING THE OCCURRENCE OF CERTAIN PLANKTON ORGANISMS IN SERIES OF STOMACHS TAKEN FROM FISH CAUGHT IN MOUNTS BAY AND MEVAGISSEY BAY FROM THE 18TH OF AUGUST, 1913, ONWARD.

Positions	South of St. Michael's Mount, 1 mile.										S.E. × S. of St. Michael's Mount 3 miles.										S. × E. of Penzance 1 mile.										S.W. × S. of S. Michael's Mount 2 miles.										Off Polperro 1 mile.									
Dates	August 18th.										August 18th.										August 19th.										August 20th.										August 27th.									
	A										B										C										D										E									
Sample Nos	1	2	3	4	5	6	7	8	9	10	11	12	13	14	15	16	17	18	19	20	21	22	23	24	25	26	27	28	29	30	31	32	33	34	35	36	37	38	39	40	41	42	43	44	45	46	47	48	49	50

(Table of plankton occurrence symbols: cc = very common, c = common, + = present, r = rare, rr = very rare, – = absent; species rows: …lausi Giesbr, …finmarchicus Gunn, …ges typicus Kröyer, …lanus elongatus Boeck, …ongicornis O. F. Müll, …ordmanni Loven, …ipunctata Quoy & Gaim, nkton principally …olenia Stoltefothii Perag.)

The following list, intended for comparison with series A B C and D of the above table, shows the plankton species observed on several positions in Mounts Bay in tow-nettings, together with the symbol of highest frequency in any one sample :—

ZOOPLANKTON.

Acartia Clausi, Giesbr.	+
Anomalocera Pattersoni, R. Temp.	r
Calanus finmarchicus, Gunn.	+
Centropages typicus, Kröyer	c
Metridia lucens, Boeck.	rr
Microsetella atlantica, Brady	rr
Oithona similis, Claus.	r
,, plumifera, Baird	rr
Pseudocalanus elongatus, Boeck ...	+
Temora longicornis, O. F. Mull ...	cc
Caradid larvae	rr
Evadne Nordmanni, Loven	+
Podon Intermedius Lilljeb.	r
Zoeas	rr
Euphysa aurata, Forbes	r

Muggiaea atlantica, Cunn.	r
Obelia nigra, E. T. Browne	r
Plutei	r
Oikopleura dioica, Fol.	rr
Fritillaria borealis	rr
Sagitta bipunctata, Quoy & Gaim	+

PHYTOPLANKTON.

Rhizosolenia Stoltefothii, Perag ...	cc
,, alata, Brightw. ...	+
,, semispina, Hens. ...	r
Ceratium fusus, Ehbg	+
,, tripos, O. F. Mull ...	+
Peridineum depressum, Ehbg ...	r
,, pallidum, Ostf.	rr
Prorocentrum micans, Ehbg	r

The following list, intended for comparison with series E. of the above table, shows the plankton species observed at several positions in Mevagissey and St. Austell's Bays in tow-nettings, together with the symbol of highest frequency in any one sample :

ZOOPLANKTON.

Acartia Clausi, Giesbr.	+
Anomalocera Pattersoni, R. Temp.	rr
Calanus finmarchicus, Gunn.... ...	+
Centropages typicus, Kröyer	cc
Corycaeus anglicus, Lubb.	rr
Oithona similis, Claus	rr
Pseudocalanus elongatus, Boeck ...	+
Paracalanus parvus, Claus.	rr
Temora longicornis, O. F. Mull ...	cc
Caradid larvae	+
Evadne Nordmanni, Loven	c
Podon Intermedius Lilljeb.	r
Zoeas	rr
Muggiaea atlantica, Cunningham...	rr
Obelia nigra, E. T. Browne	r

Phialidium temporarium, E. T. Browne	rr
Plutei	rr
Oikopleura dioica, Fol.	r
Fritillaria borealis	rr
Sagitta bipunctata, Quoy & Caim	+

PHYTOPLANKTON.

Rhizosolenia Stoltefothii, Perag ...	cc
,, alata, Brightw. ...	+
,, semispina, Hens. ...	r
Ceratium fusus, Ehbg.	+
,, tripos (O. F. Mull) ...	+
Dinophysis acuta (Ehbg)	rr
Peridineum depressum, Bail	r
Prorocentrum micans, Ehbg	rr

the touch. The single species responsible for this condition was found to be *Rhizosolenia Stoltefothii* Perag., in the profuse gatherings of which, amounting in one sample to 29 cc. of material (a five min. surface haul with the fine net), were also to be observed in comparative scarcity one other species of diatom, *Rhizosolenia alata* Bright, and certain Peridiniales.

The phytoplankton, however, did not appear to be evenly distributed throughout the whole area, but occurred in greater abundance in patches toward the centre and eastern outer quarter of the Bay. At one position about half a mile S.E. × S. of St. Michaels Mount, a slight but comparatively clean sample of zooplankton was obtained, but about three miles farther to S.E. of this station phytoplankton again

occurred in quantity. The zooplankton samples taken at
several positions were not large, but exhibited a variety of
species, including certain oceanic Copepods, *e.g., Oithona
plumifera, Microsetella atlantica.* etc., in sparing numbers.
The bulk of such tow-nettings were made up by two species
of Copepods, *Centropages typicus* and *Temora longicornis*,
but it was a noticeable fact that where the heavier gatherings
of phytoplankton were taken the quantity of zooplankton was
slighter in proportion.[19]

It is now a well-known fact that throughout the whole of
August the Mounts Bay fishery was practically non-existent
for, as may be gleaned from the pages of the " Fish Trades
Gazette," the majority of the local drifters were catching only
a few hundred fish per night; in fact during the short period
in which material for the above series of observations were
gathered it was occasionally a difficult matter to obtain even a
sufficient supply of fish for examination.

It will be seen that the majority of the stomach samples
examined exhibited food essentially of a vegetable character
with a slight admixture of animal matter, a few showed layers
consisting of phyto- and zooplankton respectively, not sharply
defined it is true, but sufficiently well marked to be determined
by a difference in colour, whilst a minority were composed
almost entirely of zooplankton with but a slight admixture of
vegetable matter.

Much valuable evidence is afforded by this series of observations, when considered in comparison with the tow-nettings,
in support of a theory that the extent of migration is determined
to a large extent by the condition of the food supply, but this
is a question beyond the province of the present paper. It
certainly appears that phytoplankton does not constitute a
type of food selected by choice, otherwise Mounts Bay would
have proved an ideal feeding ground.

At the middle of September of the present year Mr.
Howard Dunn of Mevagissey informed us that incoming
vessels reported large masses of pilchards congregated in mid-

---

[19] A possible explanation of this feature may lie in the fact of the rapid
clogging of the nets at these stations, referred to above.

Channel, from which it would appear that although the fish were present on the surface no serious migrations to the inshore feeding grounds had as yet taken place.

A final observation, for which we are indebted to Mr. Matthias Dunn of Newlyn, is therefore one of considerable interest. This observer, in a letter dated October 3rd, states: "We had a unique experience on Monday of this week. Just off Mousehole there were evidences of large quantities of fish, gannets were falling in numbers, and finally we saw a basking shark. When we came directly on to the ground we observed some crustaceans forming themselves into long furrows with knots here and there. The shark appeared to be feeding on these organisms, together with a large number of pilchards and small mackerel. We finally shot our nets with the result that about eleven thousand fish were taken. I took out the contents of the stomach of one of the mackerel, to find some well preserved specimens of crustacea."

"The next day we went out again to find the water of a dark olive green hue shot with red, and in standing in toward the coast found that this colour terminated quite suddenly, leaving the water, with its burden of crustaceans, quite distinct. One could tell within a yard where the coloured water terminated and the other type began."

"A day or two after this one of the sharks, whilst harassing a shoal of pilchard, rose suddenly outside them and drove them into one of the coves, where they were taken in a seine and realized when marketed £1,200. A few weeks prior to this a catch of seined fish realized £500. So you will understand that the fishery is improving."

The sample referred to consisted of one stomach contents of a mackerel (not the stomach itself) contained in a glass tube, and obviously it was impossible to determine whether the component species were distributed in layers. Upon examination, however, these were found to consist almost entirely of the two Copepods, *Anomalocera pattersoni* and *Calanus finmarchicus*.

In a letter to the Editor of "Nature"[20] we recorded a visitation of the former species to Mounts Bay during the first

20 "Nature" No. 2279, Vol. 91, p. 451.

week in June of the present year. The tow-nettings taken in August did not show the species other than in comparative scarcity at any of the stations. It is fair to suppose therefore that the shoal of *Anomalocera* and *Calanus* observed by Mr. Matthias Dunn formed a later immigration into the coastal waters.

In a paper upon the mackerel,[21] previously referred to, mention is made of certain types of water described by the fishermen as being of a " yellow," " blue " or " green " colour. We already possess sufficient evidence in the form of stomach material taken from mackerel caught in " yellow " water to show that this colour is probably due entirely to the presence of a vast shoal of Calanoids, not necessarily of one species, congregated at the surface. Mr. Dunn's observation appears to throw some light upon the two other types of water, which are considered by fishermen to offer attraction to drift fishes,[22] for if we may rely upon the evidence afforded by a single sample, it would appear probable that a green or blue colour is imparted to the surface of the water under varying conditions of light by the presence of a dense shoal of *Anomalocera*, which, it is well known, is of a vivid emerald hue. The suggestion of yellow referred to in the above letter might possibly have been due to segregated shoals of *Calanus*.

The most important consideration, however, afforded by these Mounts Bay observations generally is that there is some ground for supposing that whilst phytoplankton remained throughout the area in profusion, the fish were not attracted to the coastal waters, whereas on the first advent of shoals of certain zooplankton organisms, which were undoubtedly of high nutrient value, the fish began to arrive in their usual summer haunts, and the fishery became productive. From which we may presume that zooplankton forms " food of preference " as far as the pilchard is concerned.

Passing now to a consideration of the feeding habits of the pilchard, it may be remarked that there is a certain amount of

---

21 "Plankton Studies in relation to the Western Mackerel Fishery," p. 290.
22 *Id.*, p. 288.

evidence to show that this fish does not feed indiscriminately upon any form of the plankton it may happen to encounter.

The view advanced by previous authors upon general ichthyology that certain surface feeding fishes, *e.g.*, mackerel and some members of the *Clupeidae*, obtain their food supply by means of a process of filtration, has given rise to a more or less general belief that the action of feeding is a passive and involuntary one.

In the case of herring and mackerel[23] frequenting the English Channel it has already been shown that a certain capability for selective feeding does exist, which may extend to organisms of minute size. But since undue generalization upon the habits of nearly allied species of fish affords no reliable basis of argument, we have to consider those observations alone bearing relation to the pilchard.

Perhaps the most valuable observation in support of any theory of this selective capability being exercised by pilchards in feeding is that made by Cunningham[24] upon larval fish reared at Plymouth. This author states : " On the fifth day (after hatching) I put some of the minute creatures gathered from the sea by the tow-net into the tank containing the larvae, and also some minute particles of minced sea worms, and they began to feed. When five days old the larvae . . . were seen to peek or strike at the particles, which they swallowed. So that feeding is a deliberate and active, not a passive involuntary process." It is only fair to suppose that if such a capability for selective feeding were known to be exercised by any species of fish at such an early age, it would be developed the more strongly in the later stages of its growth. The evidence which we at present possess in support of such a theory so far as it affects the adult fish, may be regarded as negative.

It has already been shown that throughout the present investigation it has not been found possible in any instance to determine the presence of the same number of plankton species in a stomach as that occurring in tow-nettings taken on the same position from whence the fish were derived. Generally,

---

[23] Bullen, G. E., " Some notes upon the Feeding Habits of Mackerel and certain Clupeoids in the English Channel " (Journ. M.B.A., Vol. ix., No. 3).

[24] Cunningham, J.P., " Mark. Mar. Fishes," p. 173.

however, the species observed in the one were common to the other, not necessarily in the same relative paucity or abundance; but in certain cases food of a different character to that exhibited by the plankton samples taken from the same area, has been found in stomach contents.

Again, it has been seen that in common with the Channel and Irish mackerel[25] occasional samples have been observed in which certain species of plankton organisms have been distributed in sharply defined layers.

Moreover, in 1906, when, as may be seen from the plankton tables published in the International Bulletins, phytoplankton occurred in high preponderance over zooplankton at those stations adjacent to the fishing area throughout the summer, the majority of the stomach samples were composed solely of animal matter.

Further, it may now be stated that the same conditions as those described in the case of mackerel from the Western Fishery[26] have been consistently observed in the stomachs of pilchards, viz., that where the contents consisted largely or wholly of zooplankton the stomach walls were very thin, the organ itself being capable of considerable expansion, whereas in those cases where phytoplankton formed the bulk of the food material, the stomach walls were of moderate thickness, and the cavity of small size, incapable of being enlarged under pressure. It follows therefore that, as we have invariably observed to be the case, a far greater bulk of material is to be found in the stomach of a pilchard which had been recently feeding on pure zooplankton than in any in which phytoplankton alone was in evidence.

The majority of fishermen are agreed in stating that a distinct movement is undertaken at nightfall by pilchards which have been feeding during the daytime close inshore. Mr. Howard Dunn informed the writers that, in his opinion, this outward movement takes place even when they are situated at some distance from the coast. The same observer suggested that this might be prompted by a desire on the part

---

[25] *Vide* Bullen, " Plankton Studies, etc.," and Farran, " Rep. on Sea and Inland Fisheries, Ireland," 1901, Pt. II., p. 122.

[26] Bullen, G. E., " Some notes upon the Feeding Habits, etc.," p. 395.

of the pilchard to shoal more densely during the night and thus to minimise the danger of attack from predatory fishes.

Fishermen generally state that dogfish will not enter into the midst of a close shoal, owing to the pollution of the water by the excrement of the densely packed mass, but prefer to harass the flank, carrying on as it were a kind of guerilla warfare, and it is possible therefore that in venturing almost upon the shore at day-time, as the pilchard undoubtedly does during the height of inshore migration, the fish feel a certain amount of security from their natural enemies.

That a return to deeper water at dusk certainly does take place is evidenced by the fact that when the drift nets are shot parallel to the shore the catch is invariably taken on the one side of the fleet.

The present authors, on certain occasions when the fishery was being carried out close to the coast line, have observed the actual movement of a shoal coming off the shore. The experience, however, is by no means a common one, and fishermen agree in stating that a good catch can generally be anticipated when the occurrence is sufficiently evident to be seen and heard.

So far as our personal observation is concerned, it may be stated that at the time of movement we have seen the surface of the sea presenting an appearance as if fine rain was falling upon it. A peculiar rustling sound, similar to the wind disturbance of fallen leaves, accompanied this visual impression, but this is seldom sufficiently pronounced to be detected by the ordinary observer, though many experienced fishermen assert that they can " hear the fish coming " even when the light is insufficient for them to see the appearance of the surface of the water. In the course of our personal experience we have never found such a statement, when made by a fisherman, to be incorrect if we may take the extent of a catch as evidence. It may be stated also at this point that many fishermen are able to detect the presence of pilchard by their sense of smell, and this is quite credible when we consider the peculiar rankness of the odour emitted from the freshly caught fish.

But whatever speculative views we may entertain with regard to the actuating cause of this supposed movement off the shore, one very important possibility has to be taken into consideration, namely, that of the vertical, diurnal and nocturnal distribution of those plankton organisms which, as has been shown, constitute the principal food of the pilchard.

In those areas or stations on the Cornish seaboard from whence these observations have been derived, unfortunately up to the present no plankton samples have been taken with a view to showing the bathymetrical rise and fall of certain individual species.

We have ourselves, however, taken sufficient observations[27] to show that the majority of Calanoid Copepods materially lessened in quantity in surface tow-nettings taken within a close proximity to the shore in the one area, viz., Mevagissey and St. Austell's Bays, in which we have made an intensive (diurnal) planktonic survey. From this we may adduce some evidence in support of a theory that the richer feeding grounds do not lie close to the shore, and that pilchards at dusk seek the deeper water for feeding purposes.

Again, although we have personally had but few opportunities of examining fish taken in daylight, we have often heard it stated by fish curers and others that seined fish appear to have little or no food in the stomach. This is certainly slender evidence in support of any theory that pilchards feed more readily by night, but when taken in association with other considerations, some importance has necessarily to be given to it.

For it has been determined that in the case of certain Calanoids, notably *Calanus finmarchicus*, the maximum surface distribution occurs between the late evening and early morning.[28] Now this condition coincides in the matter of time with the pilchard's movement into deeper water, to which we have already referred. We thus have some evidence to show that this off shore migration is actuated by a desire on the part of the fish to seek the most suitable feeding

[27] Note.—The full results of these observations will be embodied in a later paper.
[28] Esterley, C. O., "The Occurrence and Vertical distribution of the Copepoda of the San Diego Region, with particular reference to nineteen Species," Univ. of California publications in Zoology, Vol. ix., No. 6, and certain other papers.

grounds. The apparent hurry in which the fish leave their diurnal haunts would seem moreover to indicate a possibility that the pilchard does not enter upon the process of feeding to any large extent until night time, and that it is impelled by hunger to travel hastily into deeper water.

The evidence afforded by the condition of the stomach contents tends in every way to support this theory. For it may be stated that when fish are dissected immediately upon coming to deck, and the contents of the stomach washed out into preservative, it is possible in the majority of cases to find the component material having undergone no change. The Copepods, in particular, are often to be found in practically intact condition.

From the above considerations generally it is possible to draw the following deductions :—

(1)   That the food of the pilchard, in coastal waters, con-sists very largely of plankton.

(2)   That from the food standpoint plankton may be divided into three distinct groups.

To the first we may ascribe those zooplankton organisms which are undoubtedly of high nutrient value, e.g., the Calanoid Copepods enumerated in the tables, together with certain developmental forms of the higher crustacea, e.g., Zoeas, Caradid larvae, etc.

To the second all phytoplankton, with the exception per-haps of the indefinable material known as " the spores of the olive seaweed," together with certain Metazoa, e.g., Oikopleura dioica, invariably associated with phytoplankton in stomach contents. Such material would appear to rank as food of a secondary character.

In the third division can be placed such plankton organisms as Medusae, which have not yet been found to occur in stomach samples.

(3)   That food material of the first group, i.e., zooplankton, is preferred to that of the second, i.e., phytoplankton, and that the organisms placed in the third group are avoided, not only as probably offering no nutrient value whatever, but by reason of the possession of irritant or offensive organs, rendering them distinctly obnoxious.

(4)   That in the avoidance of certain plankton species and in the taking of others, pilchards in their adult condition maintain the same selective, deliberate method of feeding. observed in them during their early developmental stages.

(5)   That there is a certain amount of evidence to show that feeding is largely undertaken at nightfall, when the surface distribution of certain highly nutrient plankton species reaches a maximum.

(6)   That it is largely an inherent instinct on the part of the fish to seek the richest feeding grounds, that impels them at times to hurriedly leave the protection of the shore at dusk and to move into deeper water.

Printed by BoD™in Norderstedt, Germany